Sara Blasco Hernanz
Adrián Duro Peinó
Jorge Lozano Mendoza

Distribution and abundance of martens in the Community of Madrid

Sara Blasco Hernanz
Adrián Duro Peinó
Jorge Lozano Mendoza

Distribution and abundance of martens in the Community of Madrid

Anthropic and natural determinants

ScienciaScripts

Imprint

Cover image: www.ingimage.com

This book is a translation from the original published under ISBN 978-620-2-24476-3.

Publisher:
Sciencia Scripts
is a trademark of
Dodo Books Indian Ocean Ltd. and OmniScriptum S.R.L publishing group

120 High Road, East Finchley, London, N2 9ED, United Kingdom
Str. Armeneasca 28/1, office 1, Chisinau MD-2012, Republic of Moldova, Europe
Managing Directors: Ieva Konstantinova, Victoria Ursu
info@omniscriptum.com

Printed at: see last page
ISBN: 978-620-8-50524-0

Index

Summary

Habitat selection is a key aspect of animal life and activity. Knowledge of their ecological requirements allows a better understanding of their biology, as well as a better management and mitigation of the conservation threats they face. This study evaluates the importance of different anthropic and natural variables for the distribution and abundance of the marten *(Martes foina*, Erxleben, 1777) in the Autonomous Community of Madrid. For this purpose, 90 transects of 1 km were sampled and relative abundance indices were calculated based on the frequency of occurrence of droppings. Data on the presence of different species were collected in the field, and values of different anthropic and natural variables at two spatial scales (territory and landscape) were calculated on GIS layers. With the presence data and climatic variables, a distribution model (SDM) was obtained, which indicated that the ideal habitat for martens is concentrated in the north and southwest of the region. General linear models (GLM) of abundance showed a negative relationship of marten abundance with anthropic factors (urbanization, linear infrastructure, livestock and agriculture), while it was positive with some natural variables (shrub and water cover, and bobcat abundance), and negative with others (grass cover).

The importance of marten conservation is emphasized because of its role in ecosystems, including as a seed disperser or predator (regulation of prey populations such as rodents). Thus, measures related to the maintenance of its favorable conservation status should be carried out to ensure the persistence of a species that provides important ecosystem services.

Key words: distribution, anthropogenic factors, natural factors, marten, habitat, abundance indices, indirect signs, MaxEnt.

Introduction

Since 1970 the abundance of terrestrial vertebrates has declined by 60%, which is at least 100 times more than would be expected naturally (Gonçalves-Souza *et al.*, 2020). The most prominent cause of these extinctions is the alteration of ecosystems mainly due to agriculture (Gonçalves-Souza *et al.*, 2020). However, it is not the only activity that directly influences biodiversity loss; there are others such as extractive industries, energy production, water management, logging, other land use changes, pollution and climate change among others (Kok *et al.*, 2018).

As a consequence, many landscapes have been transformed, giving rise to semi-natural mosaics that present different levels of alteration intensity, mixed with patches of the original habitats. The magnitude of their effect on biodiversity will depend on the ability of species to adapt to these new mosaics (Semenchuk *et al.*, 2022).

Mesocarnivores such as the marten (*Martes foina*) play a key role in ecosystems by providing ecosystem services as predators, competitors, umbrella species, seed dispersers, scavengers, among others (Recio *et al.*, 2015) and changes in their abundance or in the composition of their communities have repercussions on ecosystem alterations and their functionality (Recio *et al.*, 2015). However, in most of the literature they appear described by the negative impact they produce in agricultural systems, or even in natural habitats, such as predation of endangered species or those of hunting value, crop damage or monopolization of habitat resources (Ćirović *et al.*, 2016; Lozano *et al.*, 2019).

The marten is a medium sized mustelid with characteristic morphological features of nocturnal habits and adaptation to arboreal environments (Goszczyński *et al.*, 2007). It has a Palaearctic origin, occupies most of central and southern Europe and a strip in central Asia (Virgós & García, 2002). In Spain it is distributed throughout the peninsular territory, but is unevenly distributed; it is locally abundant in some areas and rare or absent in large, apparently optimal areas (Mangas & Rey Juan Carlos, 2017).

The habitat requirements of the marten have been described as non-specific, i.e., it is found in a wide range of conditions from diverse forests (Mediterranean grasslands, deciduous and coniferous), urbanized areas, rocky areas or open areas (steppes and crops), varying according to their geographic distribution (Virgós *et al.*, 2010). For example, some studies show that in the center of the Iberian Peninsula the marten has a preference for wooded habitats and rocky areas (Virgós *et al.*, 2000).

The marten is a generalist species whose diet varies seasonally (it prefers micromammals in spring-summer and fruits in autumn-winter) and according to its geographic distribution, being more carnivorous in southern regions and more frugivorous in the north (Mangas & Rey Juan Carlos, 2017). It presents a wide spectrum of prey, being able to feed from mice and shrews to rabbits, lizards, birds and insects. It is also a scavenger species and an opportunistic predator. Regarding plant matter, it consumes sweet and fleshy fruits such as junipers, blackberries, rose hips, strawberry trees and bearberries (Barrientos & Virgós, 2006).

Interactions between plants and frugivores are of utmost importance to ensure the functioning of most terrestrial ecosystems. Thus, alterations in habitat structure (e.g., habitat fragmentation), degradation of the frugivore community (e.g., defaunation) or modification of their behavior (e.g., changes in diet) can alter the functional mutualism between these groups. Consequently, new seed-dispersing interactions may emerge or, , the loss of key ecosystem linkages (Burgos *et al.*, 2024).

Mesocarnivores exhibit functionally unique characteristics compared to other specialized frugivores such as birds (Traveset *et al.*, 2014), which generally derive from their distinct dietary preferences, movement abilities, and habitat use (Burgos *et al.*, 2024). However, few studies have addressed how ecological disturbances affecting carnivore communities influence the seed dispersal process.

Altering the abundance and behavior of key frugivores, such martens, can affect the quantity (i.e., the number of seeds) and quality (i.e., the probability that a dispersed seed will survive to become a new reproductive adult) components of seed dispersal effectiveness (Burgos *et al.*, 2024).

Habitat selection is considered a key aspect of faunal activity as it affects their fitness, the choices an individual makes about which habitats to use in a given environment that reflect a balance between maximizing energy intake and corresponding constraints. Many ecological factors would positively and negatively affect habitat choice: for example, food availability, predation risk, vegetation cover or inter- and intraspecific competition for resources (Molina-Vacas *et al.*, 2012). In the specific case of predators this is a key process, as it also shapes ecological communities through predator-prey interactions (Grenier-Potvin *et al.*, 2021).

Climatic variables such as temperature and precipitation are used in habitat selection studies at global, landscape and regional scales. These variables are used to measure the availability of fundamental primary resources such as heat, water and light, among others (Philips *et al.*, 2006). Furthermore, with these data, extrapolations can be made to other situations and forecasts with respect to climate change. These climatic variables have been previously used in studies of habitat suitability in species such as marten and marten (*Martes martes martes*) in the north of the Peninsula (Vergara *et al.*, 2016), and in other carnivore species (Bai *et al.*, 2018; Kina *et al.*, 2020).

Despite its generalist character and great adaptability both habitat selection and feeding, the loss of natural habitats is one of the main threats to the conservation of the marten as it causes forests reduced to small isolated patches where the probability of survival is low. Another cause of mortality is linear infrastructures such as roads and highways that reduce the adult and dispersing juvenile population by roadkill (Abramov *et al.*, 2016). In areas such as Ibiza, the marten was introduced by man and later in the 20th century was declared extinct in that locality, mainly due to hunting for its fur. However, hunting is not currently considered a major threat to the species except in some regions of Eastern Europe as its fur has lost value (Abramov *et al.*, 2016). Be as it may, the International Union for Conservation of Nature (IUCN) currently considers the species to be of "Least Concern" (LC) (IUCN, 2015).

Thus, studying which anthropogenic and natural factors determine the distribution and abundance of martens is a crucial tool to establish appropriate management and management measures, and thus be able to maintain a good conservation status of the species and the ecosystems in which they live, maintaining the ecological processes that provide the ecosystem services from which human society benefits (Léonard *et al.*, 2019; Lozano *et al.*, 2019).

Objectives and hypotheses

The general objective of this work is to study the ecological factors that determine the distribution and abundance of the marten in the Community of Madrid, a topic that has been little studied to date. For this purpose, a species distribution model (SDM) will be elaborated to represent the habitat suitability of the marten in the Community of Madrid based on climatic factors, calculating the area of suitable habitat for the beech marten in the region. In addition, two abundance models will be obtained, one for anthropic factors and the other for natural factors, considering both spatial scales.

Thus, the following hypotheses are proposed:

- Hypothesis 1 (H1): it is expected that the most suitable areas in the Community of Madrid for martens depend mainly on temperature, being in general the most thermal areas the least suitable.
- Hypothesis 2 (H2): marten abundance will be higher in areas dominated by trees and high rock cover.
- Hypothesis 3 (H3): marten abundance will be negatively related to areas with scarce natural vegetation and highly anthropized.

Materials and Methods

Study area

The study was conducted in the Autonomous Community of Madrid (Spain), located the central part of the Iberian Peninsula (Figure 1). Its surface area covers 8,028 km2 with a population of seven million inhabitants, with a density of over 800 inhabitants/km2 (INE, 2023), making it one of the highest in the country. This region is characterized by its diversity of landscape, where you can find highly populated areas to cultivated areas, pastures and mountainous areas.

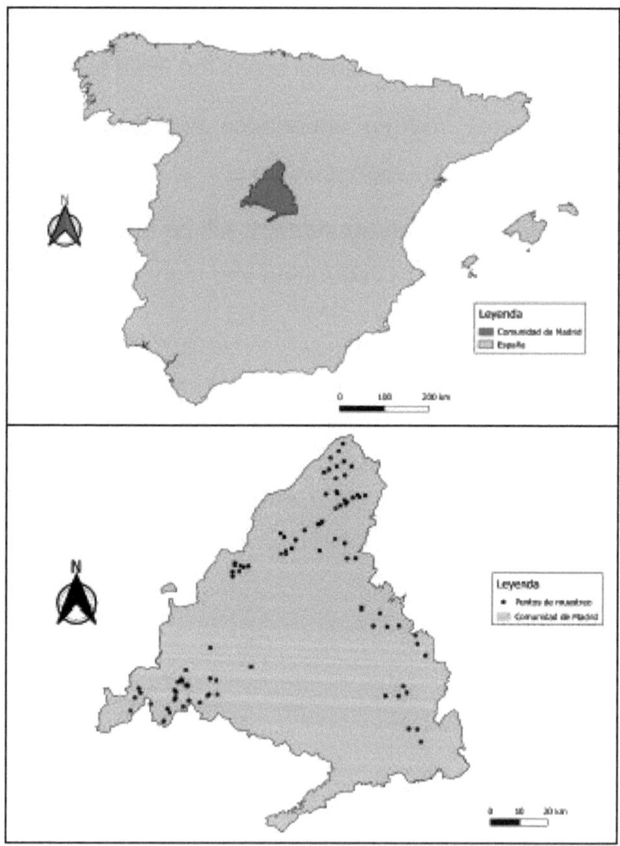

Figure 1. Distribution of sampling sites in the Community of Madrid. The black

dots represent the mean coordinate of the location of each transect.

The vegetation of the typical Mediterranean climate consists of a mixture of holm oak (*Quercus ilex ballota*) dominated forests found mainly in the southwest and center of the province, and to a lesser extent in the north and southeast. The thickets of *Cistus ladanifer* and *Retama sphaerocarpa* are the dominant elements of the undergrowth of the holm oak forests. These mosaics are interspersed with pines (*Pinus pinea* and *Pinus pinaster*) and juniper *(Juniperus oxycedrus*), which are usually accompanied by broom *(Retama sphaerocarpa* or *Salvia rosmarinus*), rockrose *(Cistus* spp.), thyme *(Thymus* spp.) or lavender *(Lavandula* spp.), among others.

Deciduous oak forests consisting of *Quercus pyrenaica* and shrubs such as *Cistus laurifolius* and *Cytisus scoparius* are also common in the mountainous areas of the territory, where the climate is usually wetter and cooler, with less pronounced drought. Pine forests at higher altitudes are formations of *Pinus sylvestris*, which resist the coldest winters and facilitate the maintenance of snow cover (Lozano, 2010).

The cultivated areas of the Community of Madrid occupy approximately 27% of its territory, being these of cereals such as barley (*Hordeum vulgare*), olive trees (*Olea europea*) and vines (*Vitis vinifera*) mainly, but also irrigated and areas intended for pasture (Directorate General of Agriculture and Livestock, 2015). The riparian vegetation is characterized by poplar trees (*Populus* spp.), elm trees (*Ulmus* spp) and various herbaceous species (Rivas- Martínez *et al.*, 1987).

Field sampling design

The methodology used to estimate the abundance of the marten consisted of an indirect sign sampling, in linear transects of 1 km in length per road, to detect the presence of martens through their depositions. Although there are various techniques to determine the abundance of a species such as direct count censuses, for this case it would not be feasible due to the fact that it is not an easily visible species, but on the contrary, it is a solitary, elusive and predominantly nocturnal animal (Mangas & Rey Juan Carlos, 2017). In addition, indirect cue sampling is efficient for martens, low invasive and low cost (Virgós *et al.*, 2010). With this methodology, an index of relative abundance can be calculated, based on the frequency of occurrence of droppings on the road, which provides a reliable approximation of the real abundance of the species (Mangas *et al.*, 2008; Martin-García *et al.*, 2022).

Marten droppings were identified by their shape, size, odor and location; when their identification was not clear, they were not considered for the study (Mangas *et al.*, 2008). Their appearance is elongated and they usually form a loop, are usually dark and very sticky, with all kinds of materials stuck to their surface (Figure 2).

Figure 2. Marten droppings on the edge of one of the sampled transects, showing their morphology, color, stickiness and size comparison with the Garmin GPS eTrex 32X.

Sampling was carried out between March and July 2022-2024, 90 transects. It is known that several carnivore species, including martens, deposit feces to mark prominent areas of their territories (Barja & Bárcena, 2002). Also, therefore, only territorial individuals are considered. The transect consisted of a length of 1km divided into 5 segments of 200m each. In each segment, in addition to marten droppings, the droppings of two other mesocarnivores were recorded: the wildcat *(Felis silvestris)* and the fox (*Vulpes vulpes*); as well as dog (*Canis familiaris*), rabbit *(Oryctolagus cuniculus)* and wild boar (*Sus scrofa*) droppings (Figure 3).

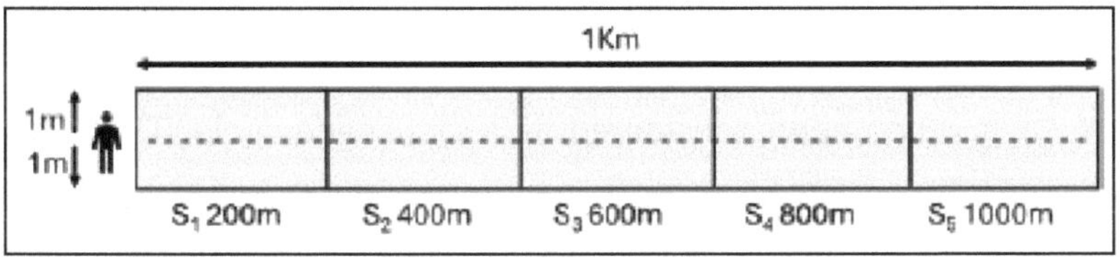

Figure 3. Schematic representation of each transect sampled.

Subsequently, indices of relative abundance of the species were calculated, based on the frequency of occurrence of signs. The number of segments in which signs of each species were present (0min-5max) was quantified ( ) and divided by the total number of segments (5). The result, thus standardized between 0 and 1, is then used as an index of relative abundance:

$$IA = \frac{N^{\underline{o}}\ de\ segmentos\ con\ presencia\ de\ indicios}{N^{\underline{o}}\ total\ de\ segmentos\ (5)}$$

To calculate the relative abundance of wild rabbits, we used the number of latrines found per kilometer following the example of Palma *et al.* (1999):

$$IA = \frac{\Sigma\ N^{\underline{o}}\ de\ letrinas\ de\ cada\ segmento}{1\ km}$$

Obtaining other data on vegetation, human infrastructure and .

Depending on the scale of study the variables affecting the abundance of a species may vary, so it is desirable to include multiple scales habitat studies (Orians & Wittenberger, 1991). Rettie & Messier (2000) proposed that at a large spatial scale the dominant pattern is avoidance of negative fitness factors, while less important limiting factors are those that influence habitat at smaller scales. Therefore, 2 spatial scales were selected to conduct the study by applying two different buffers around each transect to obtain data.

The first buffer, with a radius of 2000 m, corresponds to an area of 12.6 ^{km2} and has been considered as a landscape scale. Thus, this buffer of 1257 ha is considered to be able to gather information on the location of the marten population. The second buffer, with a radius of 750 m, corresponds to an area of 0.79 ^{km2} and is considered to cover the information of the territory of each marten (79 ha).

The marten in natural environments shows territoriality against individuals of the same sex, but they overlap their territories with those of the opposite sex, usually covering 2 ^{km2} in these environments and their area of occupancy being smaller in humanized areas (Ministry for Ecological Transition and the Demographic Challenge, 2023). The size of their home range can be a few hundred hectares and those of males can be up to three times larger than those of females. Their territories are not used homogeneously, but show a patchy pattern of occupancy depending on the availability of food and shelter (Mangas & Rey Juan Carlos, 2017).

Thus, using Geographic Information Systems (GIS), information was obtained for a total of 30 natural (e.g., forest cover) and anthropogenic (e.g., human infrastructure) variables, measured in the 2 selected buffers centered around the midpoint (600m) of each transect (see Annex, Table 1).

Species Distribution Model (SDM)

Software such as MaxEnt (Maximum Entropy) is used to represent, analyze and discuss the potential distribution of species, as well as to identify the habitats they occupy and associated factors. MaxEnt is an algorithm that is based on niche theory and predicts distribution patterns of a species in a given area solely with presence data and different environmental factors (Bai *et al.*, 2018). Thus, the input data are represented by the coordinates of the occurrence points of a species and predictors (geographic data) describing the spatial variability of the factors (biotic and/or abiotic) considered over the entire study area.

On the one hand, entropy in space is maximized, i.e., the most uniform geographic distribution of the predicted presence of the species is sought. On the other hand, similarity is maximized, i.e., an effort is made to adjust the model so that its predictions of where the species under study is found in a given geographic space are as consistent as possible with what was expected based on previous data or prior knowledge (Lissovsky & Dudov, 2021).

To generate the model, the 31 transects with marten presence were selected together with the 19 climate layers from Worldclim (Fick & Hijmans, 2017) (see Annex, Table 2) and adapted to the format required for the MaxEnt software (.csv files with presence coordinates in latitude and longitude, including the species name and environmental layer files in raster format, all layers having the same spatial resolution, extension and coordinate system). Following the example of Bai *et al.* (2018), 10 replicates were performed with "Logistic" output format, using 25% of the presence data for model fitting, while 75% of the remaining data were used to generate the final model. The rest of the settings were left as default.

The basic measure to evaluate the quality of the model generated by MaxEnt is area under the Receiver Operating Characteristic Curve (AUC_ROC). The closer the values are to 1 the more significant the model is with respect to a random model (AUC=0.5) (Phillips *et al.*, 2006); and, therefore, the better the model is at predicting habitat suitability. Values above 0.5 are acceptable models and below that threshold are rejected as invalid. In general, models with an AUC>0.75 are considered potentially useful (Elith *et al.*, 2002). In the resulting model habitat suitability was reclassified, as recommended by Chefaoui *et al.* (2005), into the following categories: very low quality (0-0.25), low quality (0.25-0.50), high quality (0.50-0.75) and very high quality (0.75-1).

Statistical analysis

The normality and homogeneity of variances were verified for the dependent variable, and if it did not meet the requirements of the parametric tests, it was normalized or its kurtosis was checked for positivity (Underwood, 1996).

Occupancy models and generalized linear models (GLMs) are also used to explore the relationship between species abundance and habitat descriptive environmental factors Bai *et al.*, 2018). Therefore, the method chosen for the analysis of quantitative data was the construction of backward stepwise generalized linear models (GLMs). Thus, 2 models were built for marten abundance, using the logarithm of the marten relative abundance index as the response variable, and the rest of the environmental or anthropogenic variables as predictors according to each model.

To reduce the number of predictor variables to orthogonal factors, Factor Analysis was performed (Osborne, 2014). It should be noted that the orthogonal factors were extracted using the principal components procedure, and rotating the axes with the "Varimax" method. Several studies have shown that using this method in statistical processes avoids multicollinearity between predictor variables (e.g. Graham, 2003).

Once the environmental variables were summarized into orthogonal factors, a GLM was obtained to study the importance of anthropogenic factors, measured at the two spatial scales indicated above, on marten abundance; and a second model was constructed with the natural factors.

Data extraction of predictor variables at the two selected spatial scales, adaptation of the raster layers for MaxEnt and calculation of the area of the different habitat suitability categories was performed with QGIS 3.34.1 for Windows. Other statistical analyses and abundance models were elaborated with Statgraphics 19 X64.

Results

Indications (i.e., droppings) of martens were found in 31 transects out of a total of 90 sampled, i.e., we obtained presence data 34.4% of the transects (with a mean value of abundance index 0.04, standard error 0.06, maximum value 0.25 and minimum value 0).

Distribution Model (SDM) for martens

The MaxEnt ROC results (Figure 4) showed a mean AUC value of 0.767, indicating that the predictions obtained from the MaxEnt model for marten presence in Madrid were of high quality or useful.

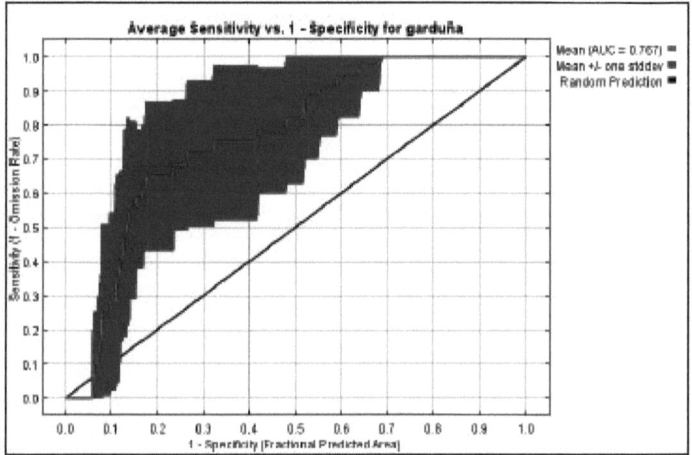

Figure 4. Representation of the ROC verification of the distribution of habitat suitability for martens in the Community of Madrid carried out by MaxEnt.

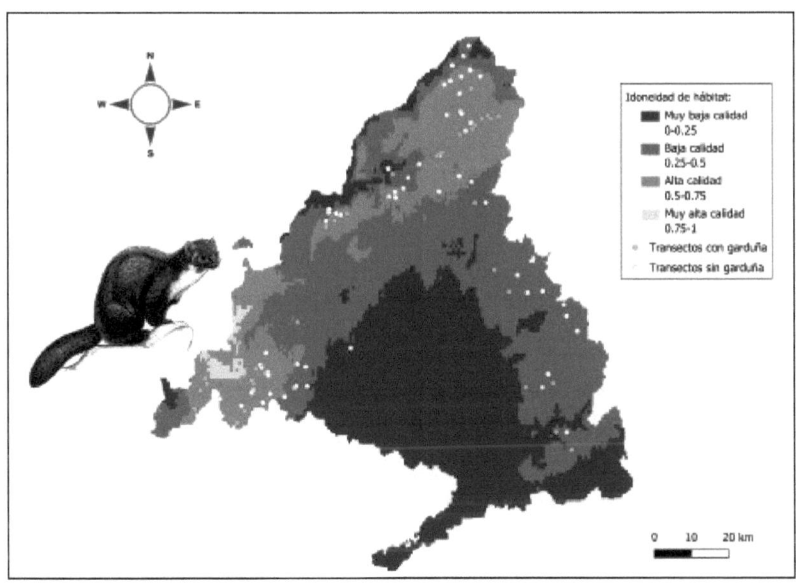

Map representing the climatic suitability of habitat for martens in the Community of Madrid with the presence data obtained in this study.

According to the distribution model obtained (Figure 5), it is estimated that the suitable or ideal habitat is located mainly along mountainous areas such as the Sierra de Guadarrama, which has an altitude between 1000-2000 m, and in the area of the marshes to the southwest of Madrid (Sierra Oeste), where we can find plains and areas of up to 1000 m. The area of high and very high quality habitat for the marten occupies an area of 1640.72 km2 (Table 1).

Table 1. Classification and quantification of the area according to its level of habitat suitability for martens in the Community of Madrid.

Categories suitability	Surface area (km2)
Very low quality	4020,69
Low quality	2353,47
High quality	1593,64
Very high quality	47,08
Total	8014,89

The results of the Jackknife estimator (see Annex, Figure S1 and Table 3) showed that seasonal variability of temperature (BIO4;

40.4%), precipitation of the coldest quarter (BIO19; 25.4%), mean temperature of the driest quarter (BIO9; 11.6%), seasonal variability of precipitation (BIO15; 9.2%) and minimum temperature of the coldest month (BIO6; 8.6%) were the five main factors contributing to the distribution (presence) of martens in Madrid.

The significance rates in the MaxEnt model prediction indicated that the minimum temperature of the coldest month (BIO6; 28.7%), precipitation of the coldest quarter (BIO19; 18.8%), mean temperature of the driest quarter (BIO9; 13.6%), seasonal variability of temperature (BIO4; 11.6%) and precipitation of the wettest month (BIO13; 9.7%) were the top five factors affecting marten distribution (see Appendix, Table 3).

Sensitivity analysis determined the influence of each factor on marten distribution (see Annex, Figure S2, S3 and S4).

Abundance model with natural variables

The Factor Analysis with natural variables at the territory and landscape scale (750m and 2000 m buffers) generated 21 orthogonal factors of which 14 were discarded for having an eigenvalue lower than 1, which indicates that the variance explained is less than that which would be explained by the variable alone (Kaiser criterion; Cuadras, 2008). The 7 orthogonal factors retained 83.98% of the accumulated variance.

Factor 1 describes landscapes with a predominance of trees and high NDVI (positive values) and areas with low grass cover (negative values). Factor 2 defines landscapes with abundant average precipitation (positive values) versus areas with low average temperature (negative values). The factor
3 describes landscapes with high grass cover (positive values). Next, factor 4 defines areas with abundant shrub cover (positive value).

Factor 5 describes rocky areas (positive values). Finally, factor 6 covers areas characterized abundant water availability (positive values) and factor 7 is characterized by areas far away from the nearest stretch of water (negative value) and high diversity of fruit crops (positive value).

Table 2. Components resulting from the Factor Analysis performed with the natural variables used to describe marten abundance (asterisks indicate significant correlation between original variables and factors, $p < 0.05$).

Variables	Factor 1	Factor 2	Factor 3	Factor 4	Factor 5	Factor 6	Factor 7
Water 2000	0,17	0,04	0,18	-0,10	-0,08	**0,81***	0,00
Water 750	0,09	-0,06	-0,12	0,04	-0,04	**0,81***	0,16
Tree 2000	**0,83***	0,12	-0,26	-0,29	-0,04	0,11	-0,02
Wooded 750	**0,76***	0,15	-0,34	-0,39	-0,02	0,14	-0,02
Crop 2000	**-0,89***	-0,14	-0,16	-0,15	-0,17	-0,13	-0,01
Crop 750	**-0,82***	-0,01	-0,16	-0,14	-0,20	-0,15	0,08
Dist. Forests	**-0,74***	0,07	-0,14	-0,09	0,06	0,04	-0,29
Dist. Water section	-0,17	0,11	-0,03	-0,03	-0,09	-0,31	**-0,66***
Scrub 2000	0,03	0,25	0,06	**0,90***	0,05	0,01	-0,02
Scrub 750	0,03	0,27	-0,06	**0,85***	0,03	-0,06	-0,06
NDVI 2000	**0,72***	0,40	0,29	0,10	-0,13	0,06	0,29
NDVI 750	**0,70***	0,40	0,30	0,10	-0,14	0,05	0,28
Pasture 2000	0,14	0,24	**0,91***	0,00	-0,04	0,00	-0,01
Pasture 750	0,07	0,23	**0,90***	0,00	-0,01	0,05	0,05
Prec Media 2000	0,06	**0,95***	0,08	0,12	0,11	-0,05	0,04
Prec Media 750	0,06	**0,95***	0,06	0,13	0,11	-0,04	0,00
Wealth fruits	0,07	0,31	0,00	-0,12	-0,04	-0,05	**0,77***
Roca 2000	0,04	0,14	-0,02	0,04	**0,94***	-0,05	0,01
Rock 750	0,03	0,12	-0,03	0,03	**0,91***	-0,06	0,02
Average Temp C° 2000	-0,22	**-0,88***	-0,27	-0,21	-0,09	-0,03	-0,12
Average Temp C° 750	-0,22	**-0,88***	0,26	-0,21	-0,09	0,03	-0,10
Eigenvalor	6,638	3,575	2,090	1,498	1,490	1,333	1,011
Percentage of Variance	31,608	17,025	9,955	7,132	7,096	6,345	4,816
Cumulative Percentage	31,608	48,633	58,587	65,719	72,815	79,161	**83,977**

Variables included in the generalized linear model, obtained using the backward stepwise regression method, between the logarithm of the marten relative abundance index and the orthogonal factors. The variance explained by the model is 26.99%.

Source	Sum of Squares	Gl	edium Square	F-Ratio	P-Value
Model	0,0988439	4	0,024711	7,86	**0,0000**
Waste	0,26733	85	0,00314506		
Total (Corr.)	0,366174	89			
Factnat_3	0,0319726	1	0,0319726	10,17	**0,002**
Factnat_4	0,0292806	1	0,0292806	9,31	**0,003**
Factnat_6	0,044524	1	0,044524	14,16	**0,0003**
Cat IA 2022	0,0151579	1	0,0151579	4,82	**0,0309**
Waste	0,26733	85	0,00314506		
Total (corrected)	0,366174	89			

These 7 orthogonal factors were used as predictor variables in the model together with the variables altitude and the abundance indices of fox, wildcat and rabbit (i.e. IA fox, IA cat and IA rabbit) in which the log IA of the marten was used as a response variable. The model obtained was highly significant. The result of the model showed that factors 3, 4 and 6, and cat AI were significantly associated with the response variable (Table 2).

Considering the results of the model (Table 3), as well as Figure 6, we can see the positive relationship between the abundance of martens (log IA marten) and the variables IA cat, factor 4 and 6, as well as the negative relationship with factor 3. The higher the grassland cover, the lower the marten abundance. Where there are more martens there also appear to be more bobcats. These results are identical at both territory and landscape scales.

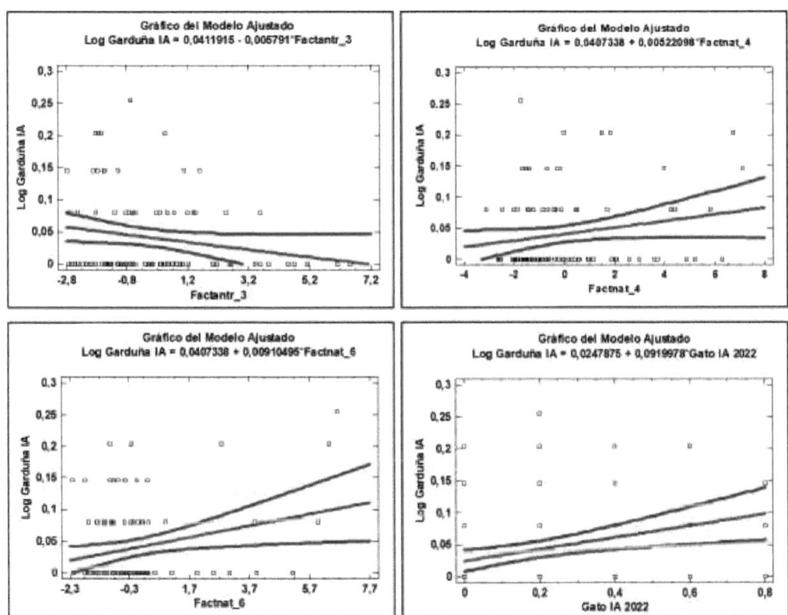

Relationship between the index of marten abundance with natural factors 3, 4, 6 and bobcat abundance.

Abundance model with anthropic variables

The Factor Analysis with territory and landscape scale variables generated 19 orthogonal factors of which 12 were discarded for having an eigenvalue of less than 1, which indicates that the variance explained is less than that which would be explained by the variable alone (Kaiser criterion; Cuadras, 2008). The 7 orthogonal factors retained 71.32% of the accumulated variance.

Table 4. Components resulting from the Factor Analysis performed with the anthropic variables used to describe marten abundance (asterisks indicate a significant correlation between the original variables and the factors, $p < 0.05$).

Variables	Factor 1	Factor 2	Factor 3	Factor 4	Factor 5	Factor 6	Factor 7
Carr/ferr. 2000	0,09	-0,36	0,27	**-0,54**	-0,30	-0,24	0,16
Carr/ferr. 750	-0,13	-0,05	-0,05	-0,28	**-0,60***	-0,10	0,49
Dist. Roads	-0,06	-0,02	0,05	-0,03	**0,74***	0,02	0,00
Dist. Hunting preserves	-0,04	-0,21	-0,11	-0,08	0,01	0,02	**-0,73***
Dist. reservoirs	-0,04	**0,87***	0,07	-0,02	-0,08	0,01	0,14
Dist. Railroad	-0,18	-0,16	0,28	**0,77***	-0,15	0,22	-0,08
Dist. Core	**-0,55***	-0,03	-0,04	0,14	0,48	0,03	0,28
Dist. Natura 2000 Network	0,00	**0,78***	-0,09	-0,25	0,03	-0,13	0,07
HH 2000 Index	**0,77***	0,29	0,11	0,13	-0,26	0,24	0,00
Index HH 750	**0,68***	0,36	-0,06	0,19	-0,43	0,19	0,08
Mixed 2000	-0,10	0,01	**0,93***	0,09	0,03	0,09	0,01
Mixed 750	0,00	-0,01	**0,93***	0,02	0,03	0,01	0,08
No. Heads Bovine	-0,07	-0,39	-0,07	-0,38	**0,46***	0,39	0,29
NºHeads Goats	0,04	-0,23	0,02	**0,77***	0,09	-0,20	0,17
No. Heads Sheep	-0,07	0,02	0,17	0,10	-0,02	**0,81***	-0,14
Number of buildings	0,21	-0,26	-0,14	-0,14	0,26	**0,58***	0,37
Dog IA	**0,43***	-0,22	-0,19	-0,18	-0,09	-0,03	0,16
Urban 2000	**0,81***	-0,26	0,01	-0,13	0,09	-0,07	-0,13
Urban 750	**0,71***	-0,01	-0,11	-0,01	0,22	-0,14	0,11
Eigenvalor	3,186	2,471	2,204	1,760	1,610	1,215	1,089
Percentage of Variance	16,766	13,007	11,598	9,263	8,473	6,392	5,732
Cumulative Percentage	16,766	29,773	41,371	50,634	59,107	65,500	**71,232**

Factor 1 describes landscapes with high human footprint values, high dog abundance and high urban land cover (positive values) and areas far from urban centers (negative values). Factor 2 was characterized by areas far from protected areas of the Natura 2000 network and reservoirs (positive values). Factor 3 describes landscapes with high mixed land cover (positive values). , factor 4 defines areas with abundant goat farming that are far from railroads (positive values) and at the same time have little land occupied by roads or railroads at the landscape scale (negative value). Factor 5 defines areas with abundant cattle ranching that are located far roads (positive values) and at the same time have little land occupied by roads or railroads at the landscape scale (negative value). Factor 6 describes areas with a high number of buildings and sheep farming (positive values); and factor 7 represents areas far away from hunting grounds (negative value).

These 7 orthogonal factors were used as predictor variables in the model together with the altitude variable, where the transformed abundance of the marten (log IA marten) was used as the response variable. The model obtained was highly significant. The model result showed that factors 3, 4 and 6 had a significant effect on the response variable (Table 4).

Variables included in the generalized linear model, obtained by the backward stepwise regression method, between the logarithm of the marten abundance index and the orthogonal anthropic factors. The variance explained by the model is 13.30%.

Source	Sum of Squares	Gl	edium Square	F-Ratio	P-Value
Model	0,0484936	3	0,0161645	4,35	**0,0067**
Waste	0,316002	85	0,00371767		
Total (Corr.)	0,364496	88			
Factnat_3	0,0185443	1	0,0185443	4,99	**0,0281**
Factnat_4	0,0251168	1	0,0251168	6,76	**0,011**
Factnat_6	0,0146603	1	0,0146603	3,94	**0,0503**
Waste	0,316002	85	0,00371767		
Total (corrected)	0,364496	88			

Considering the results of the model (Table 5), as well as Figure 7, we can see the negative relationship between marten abundance and factors 3 and 6, and the positive relationship with factor 4. Whereas, the higher the mixed land cover, the more buildings, sheep farming and land occupied by roads or railways at the landscape scale, the lower the marten abundance.

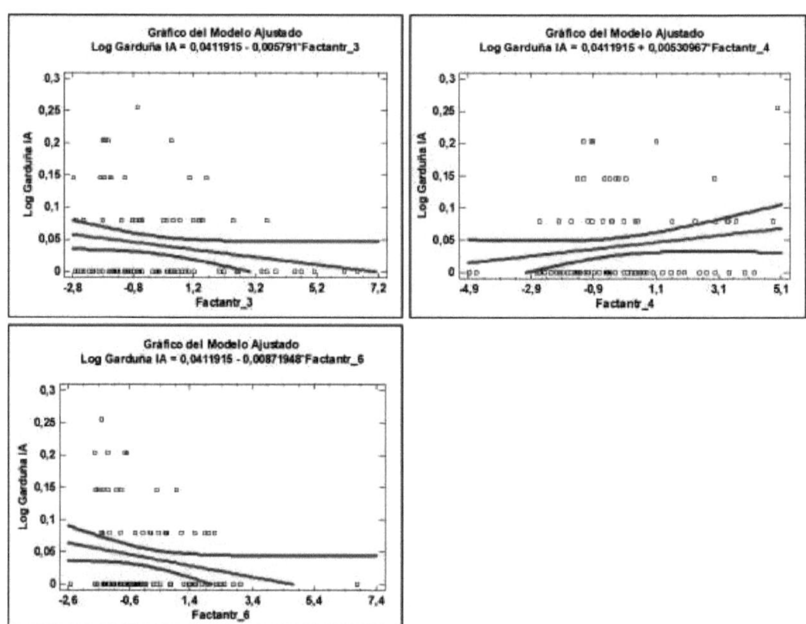

Relationship between the index of marten abundance with anthropogenic factors 3, 4 and 6.

Discussion

As a result of this study, it has been possible to increase our knowledge about the natural and anthropic factors that determine the distribution and abundance of the marten in the Community of Madrid. Regarding distribution, the results of the SDM showed that temperatures were the most determinant variables in the classification of habitat suitability, being the minimum temperature of the coldest month the most influential variable, followed by seasonal variability and the mean temperature of the driest quarter. Precipitation in the coldest quarter also played a prominent role in this selection. The importance of these zones may be due to the protection they offer against the high temperatures common in typical Mediterranean landscapes such as the plains, with abundant crops and grasslands, which are more thermal. These results coincide with our first hypothesis (H1), showing the importance of temperature in the distribution of martens.

Mustelids have a very high basal metabolic rate, on average twice as high as that of mammals of similar size, which constitutes up to 70% of their daily energy expenditure (Wereszczuk & Zaleweski, 2023). The cost of thermoregulation increases with both decreasing and increasing ambient temperature beyond the threshold of tolerance, which in martens corresponds to 25°C (Wereszczuk & Zaleweski, 2023).

According to the suitable habitat distribution map (Figure 5), the most suitable region for martens is located in the Sierra de Guadarrama. The area has a very cold Mediterranean climate and a temperate summer, with an average temperature of -0.4°C in the coldest month and 17.0°C in the warmest month, in January and July, respectively. The average annual precipitation is 1,223 mm, with a markedly Mediterranean climate, being 27 mm in July and 176 mm in November

(Sierra de Guadarrama National Park (n.d.). There is an important variety of habitats here, including Scots pine forests in the higher altitudes, holm oak and oak groves at medium altitudes, as well as scrub and meadows in open and high mountain areas (Parque Nacional de la Sierra de Guadarrama, n.d.).

In addition, the Sierra Oeste, located between the Sierra de Guadarrama and the Sierra de Gredos, is also ideal for the marten, presenting a varied relief of mountains, valleys and meadows with an extensive fluvial network with outstanding rivers such as the Alberche, and numerous streams that generate gallery forests characterized by willows, ashes and alders. Also of note are the mountain pastures, pine, chestnut, oak and holm oak forests. Overall, only 0.6% of the territory of the Community of Madrid is of very high quality and 19.90% of high quality, i.e. 79.5% of the surface area of Madrid does not appear to be suitable for martens according to the distribution model obtained.

In relation to marten abundance, it seems that natural variables have a greater influence than anthropogenic variables (27% vs. 13% of variance explained in the models). A negative relationship was observed between high grassland cover at the two scales studied and marten abundance. These landscapes do not offer refuges from predation risk and seem to be avoided (Virgós & García, 2002). In contrast, shrub and water cover favored marten abundance at both landscape and territory scales. This is probably due to the fact that the marten, besides being carnivorous, is also a frugivorous species that consumes fruits mainly in autumn (Barrientos & Virgós, 2006). It is therefore reasonable to find a higher abundance in areas where this resource is available, such as bushes of strawberry trees, blackberries, hawthorns or blackthorns, among many bushes that exist in Madrid. Many of them grow near rivers, which would explain the importance of water in the model. Likewise, the rodents that form part of the diet of martens are also associated with the shrub layer, finding more food in this habitat (Muñoz *et al.*, 2009). This would also explain why in these areas martens coincide with bobcats, which could coexist thanks to competition avoidance mechanisms (Vilella *et al.*, 2020).

The lack of a positive relationship between woodland or rocky areas and marten abundance requires the rejection of the second hypothesis (H2). In other areas of its natural Mediterranean distribution, the marten is found in wooded areas, in wooded mosaics or in rocky areas (Virgós *et al.*, 2000). Perhaps these habitats may be more used by the species in the absence of scrubland, fulfilling a similar role in terms of refuge and trophic availability.

Regarding anthropogenic factors, the results showed that high cover of disturbed, urbanized land or with a density of linear infrastructures (roads and railways), do not favor martens. Thus, the negative effect of these variables on the abundance of the species is consistent with the third hypothesis (H3). Fragmented landscapes and high road density increase mortality due to roadkill and the so-called barrier effect. In particular, martens tend to avoid unstructured areas that offer them shelter, such as grasslands and crops, seeking more natural areas in their dispersal, which makes them vulnerable when crossing roads (Fonda *et al.*, 2021). This could also explain why as distance to railroads increases, marten abundance increases.

As for the relationship found in this work between livestock grazing and marten abundance, completely unpublished, it may be complex to explain because of its predictable multifaceted character. Grazing often reduces understory vegetation and promotes the growth of certain plant species while suppressing others, which may influence the abundance of small mammals and insects that are preyed upon by martens, or the fruiting species that are also their food source (Thompson *et al.*, 2023). Thus, sheep grazing can alter the structure and composition of the vegetation, which tends to transform into a pasture, affecting the availability of food resources (Silva *et al.*, 2022) and thus harming the mustelid. On the other hand, the presence of livestock may attract other larger carnivore species, which could increase competition or predation risks for martens (Eggermann *et al.*, 2011; Linck *et al.*, 2023; Moberly *et al.*, 2003).

In addition, livestock activities frequently involve some denuding of the environment, with the construction of fences, roads and other infrastructure, which can also influence the movement and habitat use of martens (Mangas & Rey Juan Carlos, 2017). While these structures can provide shelter and burrow sites, they can also pose risks such as road mortality and habitat fragmentation (Silva *et al.*, 2022).

However, the positive relationship found between goats and marten abundance indicates that the carnivore can coexist with cattle. In more shrubby areas, where goats can easily feed because of their browsing habits (unlike sheep), if the stocking rate is not high enough to transform the vegetation, it seems, according to the results, that the mustelid can thrive without problems.

The conservation of martens is relevant due to their important ecological role, their interactions with other species and the ecosystem services they provide to . As mesocarnivores they play a vital role in controlling prey populations, mainly rodents, thus maintaining a balanced ecosystem. They are also important in seed dispersal, which helps in forest regeneration, recovery of degraded areas and restoration of biodiversity (Santos & Santos-Reis, 2010; Traveset *et al.*, 2014; Burgos *et al.*, 2024).

Particularly in Mediterranean ecosystems, martens have been observed to use a wide variety of habitats (Recio *et al*., 2015; Fonda *et al*., 2021), and the results of this work show that they are sensitive to excessive alteration of the natural environment. These characteristics make them key indicators of ecosystem health, as their presence, abundance, and behavior can reflect changes in environmental conditions and significant human impacts.

Conclusions

The findings suggest that the transformation of the landscape by human activities is a determining factor in the distribution and abundance of martens in the Community of Madrid. The anthropized, cultivated and livestock areas not only reduce the available habitat, but also alter the availability of resources. Urbanization and roads fragment the habitat, which can increase mortality and limit marten dispersal movements.

Promoting structured landscapes, with a high degree of naturalness, and maintaining connectivity between them, are management measures that will favor marten populations in the center of the Iberian Peninsula, as well as the recovery of degraded and abandoned areas, an objective to which marten activity itself can also help.

Bibliography

Abramov, A. V, Kranz, A., Herrero, J., Choudhury, A., & Maran, T. &. (2016). *Martes foina. The IUCN Red List of Threatened Species.* https://doi.org/10.2305/IUCN.UK.2016-1.RLTS.T29672A45202514.en.

Bai, D. F., Chen, P. J., Atzeni, L., Cering, L., Li, Q., & Shi, K. (2018). Assessment of habitat suitability of the snow leopard (Panthera uncia) in Qomolangma National Nature Reserve based on MaxEnt modeling. *Zoological Research*, 39(6), 373-386. https://doi.org/10.24272/j.issn.2095-8137.2018.057.

Barja, I. and Bárcena, F. (2002). The role of feces in the olfactory communication of the Wildcat. IX National and VI Iberoamerican Congress of Ethology. Madrid, 61.

Barrientos, R., & Virgós, E. (2006). Reduction of potential food interference in two sympatric carnivores by sequential use of shared resources. In *Acta Oecologica* (Vol. 30, Issue 1, pp. 107-116). https://doi.org/10.1016/j.actao.2006.02.006

Burgos, T., Escribano-Ávila, G., Fedriani, J. M., González-Varo, J. P., Illera, J. C., Cancio, I., Hernández-Hernández, J., & Virgós, E. (2024). Apex predators can structure ecosystems through trophic cascades: Linking the frugivorous behaviour and seed dispersal patterns of mesocarnivores. *Functional Ecology.* https://doi.org/10.1111/1365-2435.14559

Chefaoui et al., 2005: Chefaoui, R. M., Hortal, J., & Lobo, J. M. 2005. Potential distribution modelling, niche characterization and conservation status assessment using GIS tools: a case study of Iberian Copris species. *Biological Conservation*, 122(2), 327-338.

Ćirović, D., Penezić, A., & Krofel, M. (2016). Jackals as cleaners: Ecosystem services provided by a mesocarnivore in human-dominated landscapes. *Biological Conservation*, 199, 51-55. https://doi.org/10.1016/j.biocon.2016.04.027

Cuadras, M.C. (2008). *New methods of multivariate analysis*. CMC

Directorate General of Agriculture and Livestock *indicators on agriculture, livestock, food industry, animal protection and livestock trails* (2015).

Eggermann, J., da Costa, G. F., Guerra, A. M., Kirchner, W. H., & Petrucci-Fonseca, F. (2011). Presence of Iberian wolf (Canis lupus signatus) in relation to land cover, livestock and human influence in Portugal. *Mammalian Biology*, *76*(2), 217-221. https://doi.org/10.1016/j.mambio.2010.10.010.

Elith, J. 2002. Quantitative methods for modelling species habitat: comparative performance and an application to Australian plants. In: Ferson, S. and Burgman, M. (eds), Quantitative methods for conservation biology. *Springer*, pp. 39-58.

Fick, S. E., & Hijmans, R. J. (2017). WorldClim 2: New 1 km spatial resolution climate surfaces for global land areas. *International Journal of Climatology, 37*(12), 4302-4315. https://doi.org/10.1002/joc.5086

Fonda, F., Chiatante, G., Meriggi, A., Mustoni, A., Armanini, M., Mosini, A., Spada, A., Lombardini, M., Righetti, D., Granata, M., Capelli, E., Pontarini, R., Roux

Poignant, G., & Balestrieri, A. (2021). Spatial distribution of the pine marten (Martes martes) and stone marten (Martes foina) in the Italian Alps. *Mammalian Biology*, *101*(3), 345-356. https://doi.org/10.1007/s42991-020-. 00098-8

Gonçalves-Souza, D., Verburg, P. H., & Dobrovolski, R. (2020). Habitat loss, extinction predictability and conservation efforts in the terrestrial ecoregions. *Biological Conservation*, 246. https://doi.org/10.1016/j.biocon.2020.108579

Graham, 2003: Graham MH (2003). Confronting multicollinearity in ecological multiple regression. *Ecology* 84: 2809-2815.

Grenier-Potvin, A., Clermont, J., Gauthier, G., & Berteaux, D. (2021). Prey and habitat distribution are not enough to explain predator habitat selection: addressing intraspecific interactions, behavioural state and time. *Movement Ecology*, 9(1). https://doi.org/10.1186/s40462-021-00250-0. https://doi.org/10.1186/s40462-021-00250-0.

Ine, 2023: National Institute of Statistics. 2023. Results by Autonomous Communities. Resident population by date, sex and age (since 1971).

IUCN (2015). Abramov, A.V., Kranz, A., Herrero, J., Choudhury, A. & Maran, T. 2016. Martes foina. The IUCN Red List of Threatened Species 2016: e.T29672A45202514.https://dx.doi.org/10.2305/IUCN.UK.2016-1.RLTS.T29672A45202514.en. Accessed on 19 June 2024.

Kina, K. C., Bhumpakhpan, N., Trisurat, Y., Mainmit, N., Ghimire, K., & Subedi,
M. (2020). *Analysis of Potential Distribution of Tiger Habitat using MaxEnt in Chitwan National Park, Nepal*. Journal of Remote Sensing and GIS Association of Thailand, 21(3), 1-15.

Kok, M. T. J., Alkemade, R., Bakkenes, M., van Eerdt, M., Janse, J., Mandryk, M., Kram, T., Lazarova, T., Meijer, J., van Oorschot, M., Westhoek, H., van der Zagt, R., van der Berg, M., van der Esch, S., Prins, A. G., & van Vuuren, D. P. (2018). Pathways for agriculture and forestry to contribute to terrestrial biodiversity conservation: A global scenario-study. *Biological Conservation*, 221, 137-150. https://doi.org/10.1016/j.biocon.2018.03.003.

Léonard, A., Desbiez, J., Bodmer, R. E., & Santos, S. A. (2009). Wildlife habitat selection and sustainable resource management in a Neotropical wetland. *In International Journal of Biodiversity and Conservation* (Vol. 1, Issue 1). http://www.academicjournals.org/ijbc. http://www.academicjournals.org/ijbc.

Linck, P., Palomares, F., Negrões, N., Rossa, M., Fonseca, C., Couto, A., & Carvalho, J. (2023). Increasing homogeneity of Mediterranean landscapes limits the co-occurrence of mesocarnivores in space and time. *Landscape Ecology*, *38*(12), 3657-3673. https://doi.org/10.1007/s10980-023-01749-0

Lissovsky, A. A., & Dudov, S. V. (2021). Species-Distribution Modeling: Advantages and Limitations of Its Application. 2. MaxEnt. *Biology Bulletin Reviews*, 11(3), 265-275. https://doi.org/10.1134/s2079086421030087

Lozano, J. (2010). Habitat use by European wildcats (*Felis silvestris*) in central Spain: what is the relative importance of forest variables? *Animal Biodiversity and Conservation*, 33.2: 143-150.

Lozano, J., A. Olszańska, Z. Morales-Reyes, A. J. Castro, A. F. Malo, M.

Moleón,
A. Sánchez-Zapata, A. Cortés-Avizanda, et al. (2019). Human-carnivore relations: A systematic review. *Biological Conservation*, 237, 480-92.

Mangas, J. G., Lozano, J., Cabezas-Díaz, S., & Virgós, E. (2008). The priority value of scrubland habitats for carnivore conservation in Mediterranean ecosystems. *Biodiversity and Conservation*, 17(1), 43-51. https://doi.org/10.1007/s10531-007-9229-8. https://doi.org/10.1007/s10531-007-9229-8

Mangas, J. G., & Rey Juan Carlos, U. C. (2017). In: *Virtual Encyclopedia of Spanish Vertebrates*. http://www.vertebradosibericos.org/.

Martin-Garcia, S., Rodríguez-Recio, M., Peragón, I., Bueno, I., & Virgós, E. (2022). Comparing relative abundance models from different indices, a study case on the red fox. *Ecological Indicators*, 137.
https://doi.org/10.1016/j.ecolind.2022.108778

Ministry for Ecological Transition and the Demographic Challenge (2023). *Conservation status sheet for the marten (Martes foina)*. Retrieved from https://www.miteco.gob.es/content/dam/miteco/es/biodiversidad/temas/inventarios-nacionales/ieet_mami_martes_foina_tcm30-99820.pdf.

Moberly, R. L., White, P. C. L., Webbon, C. C., Baker, P. J., & Harris, S. (2003). Factors associated with fox (Vulpes vulpes) predation of lambs in Britain. *Wildlife Research*, *30*(3), 219-227. https://doi.org/10.1071/WR02060.

Molina-Vacas, G., Bonet-Arbolí, V., & Rodríguez-Teijeiro, J. D. (2012). Habitat selection of two medium-sized carnivores in an isolated and highly anthropogenic Mediterranean park: The importance of riverbank vegetation. *Italian Journal of Zoology*, 79(1), 128-135.
https://doi.org/10.1080/11250003.2011.620637

Muñoz, A., Bonal, R., & Díaz, M. (2009). Ungulates, rodents, shrubs: interactions in a diverse Mediterranean ecosystem. *Basic and Applied Ecology*, *10*(2), 151-160. https://doi.org/10.1016/j.baae.2008.01.003. https://doi.org/10.1016/j.baae.2008.01.003

Orians, G.H. and Wittenberger, J.F. (1991). Spatial and temporal scales in habitat selection. *American Naturalist* 137: 29-49.

Osborne, J. W. (2014). *Best practices in exloratory factor analysis*. [CreateSpace Independent Publishing Platform].

Palma L., Beja P. and Rodrigues M. (1999). The use of sighting data to analyse Iberian lynx habitat and distribution. *Journal of Applied Ecology* (36): 812-824

Sierra de Guadarrama National Park (n.d.). Climate. Retrieved June 21, 2024. 2024, from https://www.parquenacionalsierraguadarrama.es/naturaleza/clima-geologia/116-clima

Sierra de Guadarrama National Park (n.d.). *Technical data sheet of the Park*. Retrieved on 14 of June 14 from 2024, from https://www.parquenacionalsierraguadarrama.es/parque/info-pnsg/86-token-pnsg

Phillips, S.J., R.P. Anderson and R.E. Schapire. 2006. Maximum entropy modeling of species geographic distributions. *Ecological Modelling*.190:231-259

Recio, M. R., Arija, C. M., Cabezas-Díaz, S., & Virgós, E. (2015). Changes in Mediterranean mesocarnivore communities along urban and ex-urban gradients. In *Current Zoology* (Vol. 61, Issue 5). https://academic.oup.com/cz/article/61/5/793/1821076

Rettie, W.J. and Messier, F. (2000). Hierarchical habitat selection by woodland caribou: its relationship to limiting factors. *Ecography*, 23: 446-478.

Rivas-Martínez, S., Fernández-González, F. and Sánchez-Mata, D. (1987). The Central System: from the Sierra de Ayllón to Serra da Estrela. *The vegetation of Spain*: 419-451. In: Peinado, M. and Rivas-Martínez, S. (Eds.). Publicaciones Universidad de Alcalá, Madrid.

Santos, M. J., & Santos-Reis, M. (2010). Stone marten (Martes foina) habitat in a Mediterranean ecosystem: Effects of scale, sex, and interspecific interactions. *European Journal of Wildlife Research*, 56(3), 275-286. https://doi.org/10.1007/s10344-009-0317-9. https://doi.org/10.1007/s10344-009-0317-9

Semenchuk, P., Plutzar, C., Kastner, T., Matej, S., Bidoglio, G., Erb, K. H., Essl, F., Haberl, H., Wessely, J., Krausmann, F., & Dullinger, S. (2022). Relative effects of land conversion and land-use intensity on terrestrial vertebrate diversity. *Nature Communications*, 13(1). https://doi.org/10.1038/s41467-022-28245-4.

Silva, S. R., Sacarrão-Birrento, L., Almeida, M., Ribeiro, D. M., Guedes, C., Montaña, J. R. G., Pereira, A. F., Zaralis, K., Geraldo, A., Tzamaloukas, O., Cabrera, M. G., Castro, N., Argüello, A., Hernández-Castellano, L. E., Alonso-Diez, Á. J., Martín, M. J., Cal-Pereyra, L. G., Stilwell, G., & de Almeida, A. M. (2022). Extensive Sheep and Goat Production: The Role of Novel Technologies towards Sustainability and Animal Welfare. In *Animals* (Vol. 12, Issue 7). MDPI. https://doi.org/10.3390/ani12070885

Thompson, L., Rowntree, J., Windisch, W., Waters, S. M., Shalloo, L., & Manzano, P. (2023). Ecosystem management using livestock: Embracing diversity and respecting ecological principles. *Animal Frontiers*, *13*(2), 28-34. https://doi.org/10.1093/af/vfac094

Underwood, A.J. (1996). Experiments in ecology. Cambridge University Press, Cambridge.

Vergara, M., Cushman, S. A., Urra, F., & Ruiz-González, A. (2016). Shaken but not stirred: multiscale habitat suitability modeling of sympatric marten species (*Martes martes* and *Martes foina)* in the northern Iberian Peninsula. *Landscape Ecology*, *31*(6), 1241-1260. https://doi.org/10.1007/s10980-015-0307-0

Vilella, M., Ferrandiz-Rovira, M., & Sayol, F. (2020). Coexistence of predators in time: Effects of season and prey availability on species activity within a Mediterranean carnivore guild. *Ecology and Evolution*, *10*(20), 11408-11422. https://doi.org/10.1002/ece3.6778.

Virgós, E., Cabezas-Díaz, S., Mangas, J. G., & Lozano, J. (2010). Spatial distribution models in a frugivorous carnivore, the stone marten (Martes foina): Is the fleshy-fruit availability a useful predictor? *Animal Biology*, 60(4), 423-436. https://doi.org/10.1163/157075610X523297.

Virgós, E., & García, F. J. (2002). *Patch occupancy by stone martens Martes foina in fragmented landscapes of central Spain: the role of fragment size, isolation and habitat structure*. www.elsevier.com/locate/actao

Virgös, E., Recio, M. R., & Cortes, Y. (2000). International journal of mammalian biology Stone marten (Martes foina Erxleben, 1777) use of different landscape types in the mountains of central Spain marten scat was sampled by Walking a series of routes along randomly selected tracks. In *Z. Säugetierkunde* (Vol. 65). http://www.urbanfischer.de/journals/mammbiol

Wereszczuk, A., & Zalewski, A. (2023). An anthropogenic landscape reduces the influence of climate conditions and moonlight on carnivore activity. *Behavioral Ecology and Sociobiology*, *77*(5). https://doi.org/10.1007/s00265-023-03331-9

Annex

Table 1. Variables used for the realization of the general linear models at landscape scale (2000 m buffer) and territory scale (750 m buffer).

Variable name	Description
Water 750 and 2000	Percentage of land area covered by water
Woodland 750 and 2000	Percentage of ground area covered by tree cover
Crop 750 and 2000	Percentage of ground area covered by crops
Distance from forest	Distance (in meters) from the transect to the nearest forest
Distance of the water section	Distance (in meters) from the transect to the nearest watercourse
Shrublands 750 and 2000	Percentage of ground area covered by shrubland
NDVI 750 and 2000	Normalized difference vegetation index
Grassland 750 and 2000	Percentage of land area covered by grassland
Average precipitation 1970-2000	Annual precipitation (measured in mm) in the period
Fruit richness	Diversity of fruit crops within the landscape buffer
Rock 750 and 2000	Percentage of soil area covered by rock
Carr/ferr. 750 y 2000	Percentage of land area covered by road or railroad tracks.
Dist. Roads	Distance from the transect to the nearest road
Dist. Hunting grounds	Distance from the transect to the nearest hunting ground.
Dist. reservoirs	Distance from the transect to the nearest reservoir
Dist. Railroad	Distance from the transect to the nearest railroad track.
Dist. Core	Distance from the transect to the urban nucleus
Dist. Natura 2000 Network	Distance from the transect to nearest Natura 2000 Network
HH Index 750 and 2000	Human Footprint Index
Mixed 750 and 2000	Percentage of area covered by disturbed soil, semi-artificial
Number of Heads of Cattle	Number of heads of cattle registered in the Community of Madrid
NºHeads Goats	Number of head of goats registered in the Community of Madrid
NºHeads Sheep	Number of head of sheep registered in the Community of Madrid
Number of buildings	Number of buildings
Average temperature	Average annual temperature (measured in ºC) in the period
Altitude	Vertical distance between the midpoint of the transect with respect to sea level (measured in meters).
IA Cat	Index of relative abundance of wildcat
AI Fox	Fox relative abundance index
AI Rabbit	Rabbit relative abundance index

Table 2. List of climatic variables used in this study.

Variable	Nombre	Traducción	Descripción
BIO1	Annual Mean Temperature	Tº anual media	
BIO2	Mean Diurnal Range (Mean of monthly (max temp - min temp))	Rango diurno medio	Diferencia de Tº entre el punto más bajo y el más alto de un día
BIO3	Isothermality (BIO2/BIO7) (×100)	Isotermalidad (BIO2/BIO7) (×100)	Índice de variabilidad de Tº. Mide lo que cambia la Tº diariamente.
BIO4	Temperature Seasonality (standard deviation ×100)	Variabilidad estacional de la temperatura (desviación estándar ×100)	Mide la amplitud de las fluctuaciones de temperatura entre las estaciones.
BIO5	Max Temperature of Warmest Month	Tº Máx. Del mes más caluroso	
BIO6	Min Temperature of Coldest Month	Tº mín. Del mes más frío	
BIO7	Temperature Annual Range (BIO5-BIO6)	Rango Tº annual	
BIO8	Mean Temperature of Wettest Quarter	Tº media del trimestre más húmedo	
BIO9	Mean Temperature of Driest Quarter	Tº media del trimestre más seco	
BIO10	Mean Temperature of Warmest Quarter	Tº media del trimestre más caluroso	
BIO11	Mean Temperature of Coldest Quarter	Tº media del trimestre más frío	
BIO12	Annual Precipitation	Precipitación anual	
BIO13	Precipitation of Wettest Month	Precipitación del mes más húmedo	
BIO14	Precipitation of Driest Month	Precipitación del mes más seco	
BIO15	Precipitation Seasonality (Coefficient of Variation)	Variabilidad estacional de precipitación	Mide la amplitud de las fluctuaciones de precipitación entre las estaciones
BIO16	Precipitation of Wettest Quarter	Precipitación del trimestre más húmedo	
BIO17	Precipitation of Driest Quarter	Precipitación del trimestre más seco	
BIO18	Precipitation of Warmest Quarter	Precipitación del trimestre más caluroso	
BIO19	Precipitation of Coldest Quarter	Precipitación del trimestre más frío	

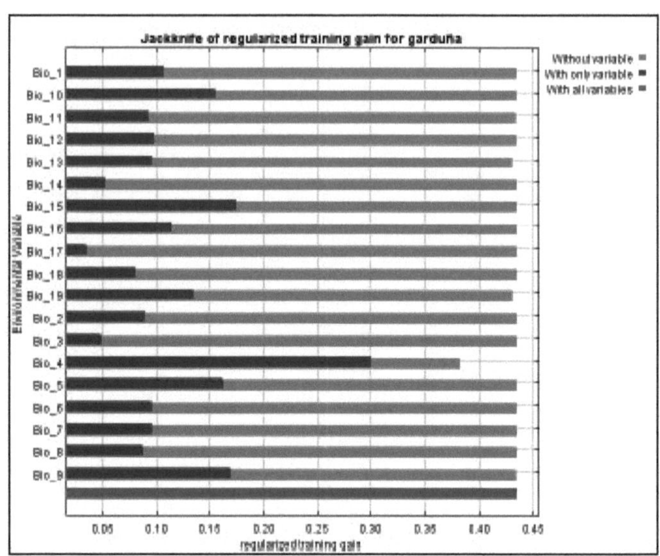

Figure S1. Jackknife test result with climate variables and training data by MaxEnt.

Results of the MaxEnt analysis showing the contribution and importance of the climatic variables used in the distribution model.

Variable	**Contribution (%)**	**Importance permutation (%)**
Bio_1 Average annual temperature	0,4	0,8
Bio_2 Average diurnal range	0	0
Bio_3 Isothermality	0	0,2
Bio_4 Seasonal temperature variability	**40,4**	**11,6**
Bio_5 Max. temp. of the hottest month	0	0
Bio_6 Min. temp. of coldest month	**8,6**	**28,7**
Bio_7 Annual Tº range	1,7	10
Bio_8 Average Tº of the wettest quarter	0	0
Bio_9 Average temperature of the driest quarter	**11,6**	**13,6**
Bio_10 **Average temperature of the hottest quarter**	0,3	0
Bio_11 Average temperature of the coldest quarter	0,3	1
Bio_12 Annual precipitation	0	0
Bio_13 Precipitation of the wettest month	**1,8**	**9,7**
Bio_14 Precipitation of the driest month	0,2	0,8
Bio_15 Seasonal variability of precipitation	**9,2**	**0,7**
Bio_16 Precipitation in wettest quarter	0,2	4
Bio_17 Precipitation in driest quarter	0	0
Bio_18 Precipitation of the hottest quarter	0	0

Bio_19 Precipitation of the coldest quarter	**25,4**	**18,8**

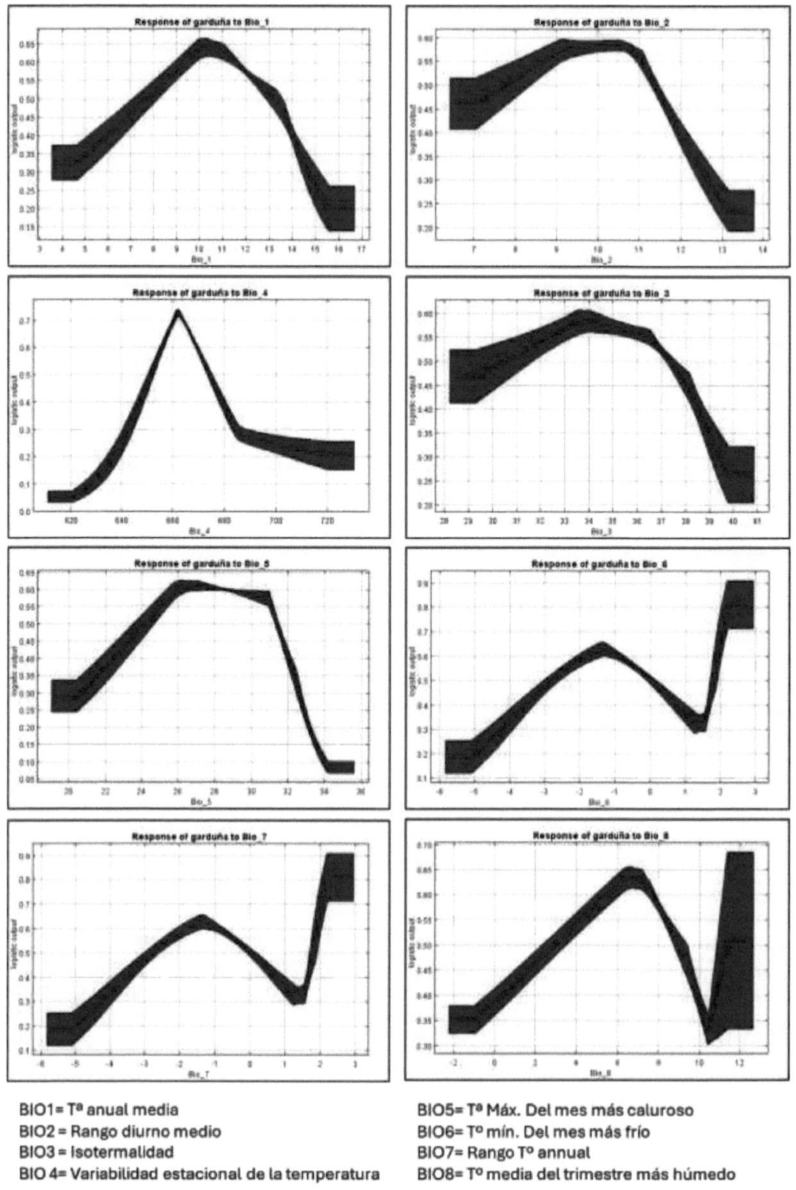

Figure S2. Response curve of selected climatic variables marten habitat suitability in the Community of Madrid.

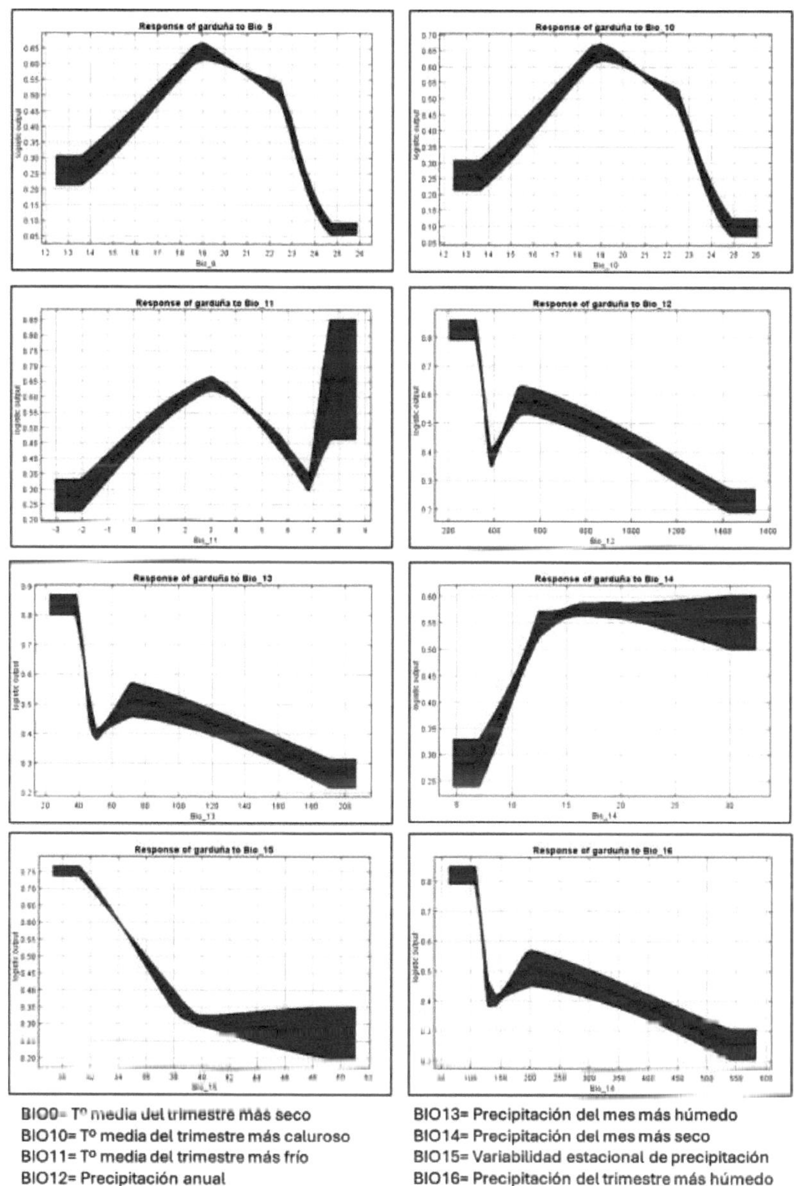

Figure S3. Response curve of selected climatic variables marten habitat suitability in the Community of Madrid.

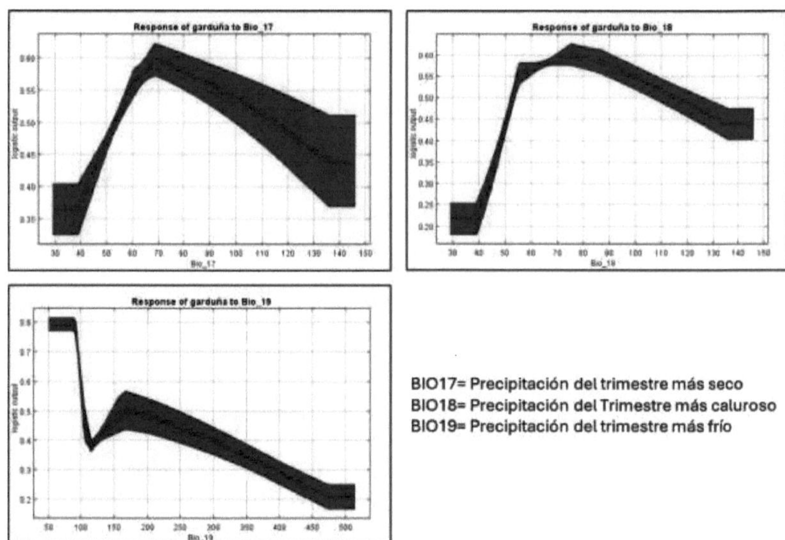

Figure S4. Response curve of selected climatic variables marten habitat suitability in the Community of Madrid.

Printed by Books on Demand GmbH, Norderstedt / Germany